W9-ADL-344
0348116 9

HEALTHY ME

Vegetables

HEALTHY ME

Published by Smart Apple Media
1980 Lookout Drive, North Mankato, Minnesota 56003

PHOTOGRAPHS BY Getty Images, Heartland Images (Paul T. McMahon), Sally Myers, The National Pork Board, Bonnie Sue Rauch, Tom Stack & Associates (Sharon Gerig, Inga Spence), Unicorn Stock Photos (Eric R. Berndt, Tommy Dodson, Mark E. Gibson, Gary Randall, Gerry Schnieders, Jim Shippee)
DESIGN BY Evansday Design

Library of Congress Cataloging-in-Publication Data
Kalz, Jill.
Vegetables / by Jill Kalz.
p. cm. — (Healthy me)
Summary: Describes different kinds of vegetables and their importance in a healthy diet.
Includes a recipe for a cup of vegetables.
ISBN 1-58340-300-0
1. Vegetables—Juvenile literature. [1. Vegetables. 2. Nutrition.] I. Title.

TX557.K25 2003
641.3'5—dc21 2002034880

First Edition

9 8 7 6 5 4 3 2 1

Vegetables

Gardens and Farms

4

Some people plant vegetable gardens. They grow just enough food for their families to eat. Other people grow vegetables to sell. Most of the vegetables we eat come from huge farms.

Some vegetables grow above the ground. Others grow below it. You can see lettuce growing in the fields. You can see tomatoes, too. But you cannot see carrots. Carrots grow below the ground. So do potatoes. You see only their green leaves.

Carrots grow down into the ground.

Farmers plant vegetables in long rows.

Farmers **harvest** vegetables with machines. Some vegetables are sent to stores right away. The rest are sent to factories. There, the vegetables are frozen or put into cans. Or they may be made into other foods.

More than half of all the vegetables grown in the United States come from California.

This farmer harvests his peppers by hand.

Kinds of Vegetables

A plant has many parts. It has **roots**, a **stem**, flowers, leaves, and seeds. The vegetables you eat come from different parts of plants. Carrots are the roots of carrot plants. Potatoes are the stems of potato plants. When you eat broccoli, you are eating the flower of the broccoli plant.

Lettuce and spinach are called leafy vegetables. You eat their leaves. Some vegetables are the seeds of plants. Peas, beans, and sweet corn are seeds.

Sweet corn can be yellow or white.

Small broccoli flowers are called florets.

Some vegetables are really fruits. Tomatoes are the fruits of tomato plants. A fruit is the part of a plant that holds seeds. Peppers and pumpkins are fruits, too. But most people use them as vegetables.

Long ago, people did not eat uncooked tomatoes. They thought tomatoes were poisonous.

Most pumpkins grown in the United States are not eaten. They are used as jack-o'-lanterns.

The Good Stuff

Vegetables have **vitamins** that everybody needs. Vitamin C helps heal cuts and bruises. It fights colds. Tomatoes have a lot of vitamin C. So do broccoli and carrots.

Vegetables also have vitamin A and fiber. Vitamin A keeps hair, skin, and eyes healthy. Fiber helps move food through your body.

Potatoes have a lot of vitamin C.

Vegetables come in all shapes, sizes, and colors.

Not all foods made from vegetables are good for you. Potato chips have salt and fat. Too much salt and fat can be bad for you. Eating fresh vegetables is the best way to get your vitamins. Most frozen vegetables are good, too.

Sweet-corn seeds grow in bunches called cobs. Cobs are covered with leaves called husks.

Potato chips are not a healthy snack.

Eating Right

All foods belong to one of five food groups. Vegetables belong to the vegetables group. Foods made from milk belong to the dairy group. There are also groups for fruits, meats, and grains.

Doctors say you should eat three to five helpings of vegetables each day. A helping may be a carrot. A glass of tomato juice. Or a few spoonfuls of cooked peas.

Onions belong to the vegetables group. ^

< Your body needs all kinds of foods.

It is important to eat foods from all of the food groups. Each group has things your body needs. Eating vegetables helps keep you looking and feeling great!

Peppers can be square and sweet, or long and spicy. They can be red, green, or yellow.

Peas grow in thin, boat-shaped shells called pods. Pods keep peas safe while they grow.

Cup of Vegetables

Three kinds of vegetables make this a tasty treat!

WHAT YOU NEED

A big bell pepper, washed
A carrot, washed and peeled
Celery, washed
Vanilla yogurt
A knife

WHAT YOU DO

1. Have an adult help you cut the pepper around the middle.
2. Clean out the seeds from the bottom half. This is your cup.
3. Have an adult help you cut the top part of the pepper, the carrot, and celery into sticks.
4. Put a little yogurt in your cup. Then put the pepper, carrot, and celery sticks in it.
5. Enjoy your vegetables. And remember to eat the cup!

WORDS TO KNOW

harvest to pick or gather food

roots the part of a plant that takes food and water from the ground

stem the part of a plant from which leaves grow

vitamins things in food that keep your body healthy and growing

Read More

Gibbons, Gail. *The Pumpkin Book*. New York: Holiday House, 1999.

Llewellyn, Claire. *Peas*. New York: Scholastic Library Publishing, 1999.

Robinson, Fay. *Vegetables, Vegetables*. New York: Scholastic Library Publishing, 1995.

Explore the Web

DOLE FIVE A DAY

http://www.dole5aday.com

F&V (FRUITS & VEGETABLES) FOR ME

http://www.fandvforme.com.au

THE PRODUCE PATCH

http://www.aboutproduce.com/producepatch

Harvesting broccoli is slow work.